Deco[illegible] a Vase

by Jodi Keller

You can decorate a vase.
Use paper and glue.
Change some of their properties.
Then, use them to make a pattern!

First, cut some paper.
You need many small pieces.
Make paper shapes!

Next, mix water with the glue.
Make a thinner liquid!

Then, paint the glue onto the vase.
Make the vase sticky!

Press the paper onto the vase.
Put one piece over another.
Make new colors!

You finished the vase.
How did you change the solids
and liquids you used?